Steck-Vaughn Shutterbug Books
Social Studies

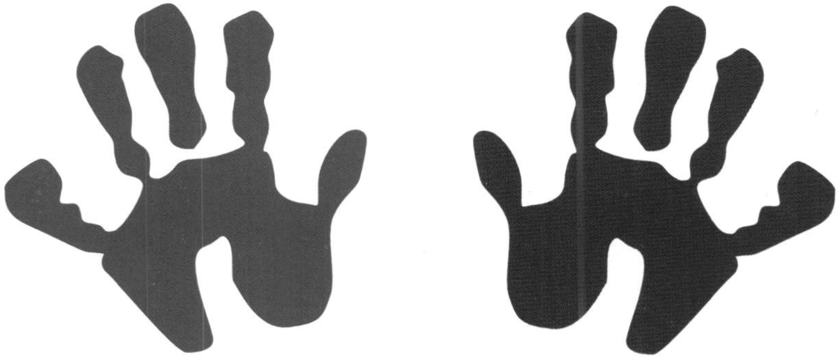

ART FOR YOU

by Ellen Catala

STECK-VAUGHN
A Harcourt Company

www.steck-vaughn.com

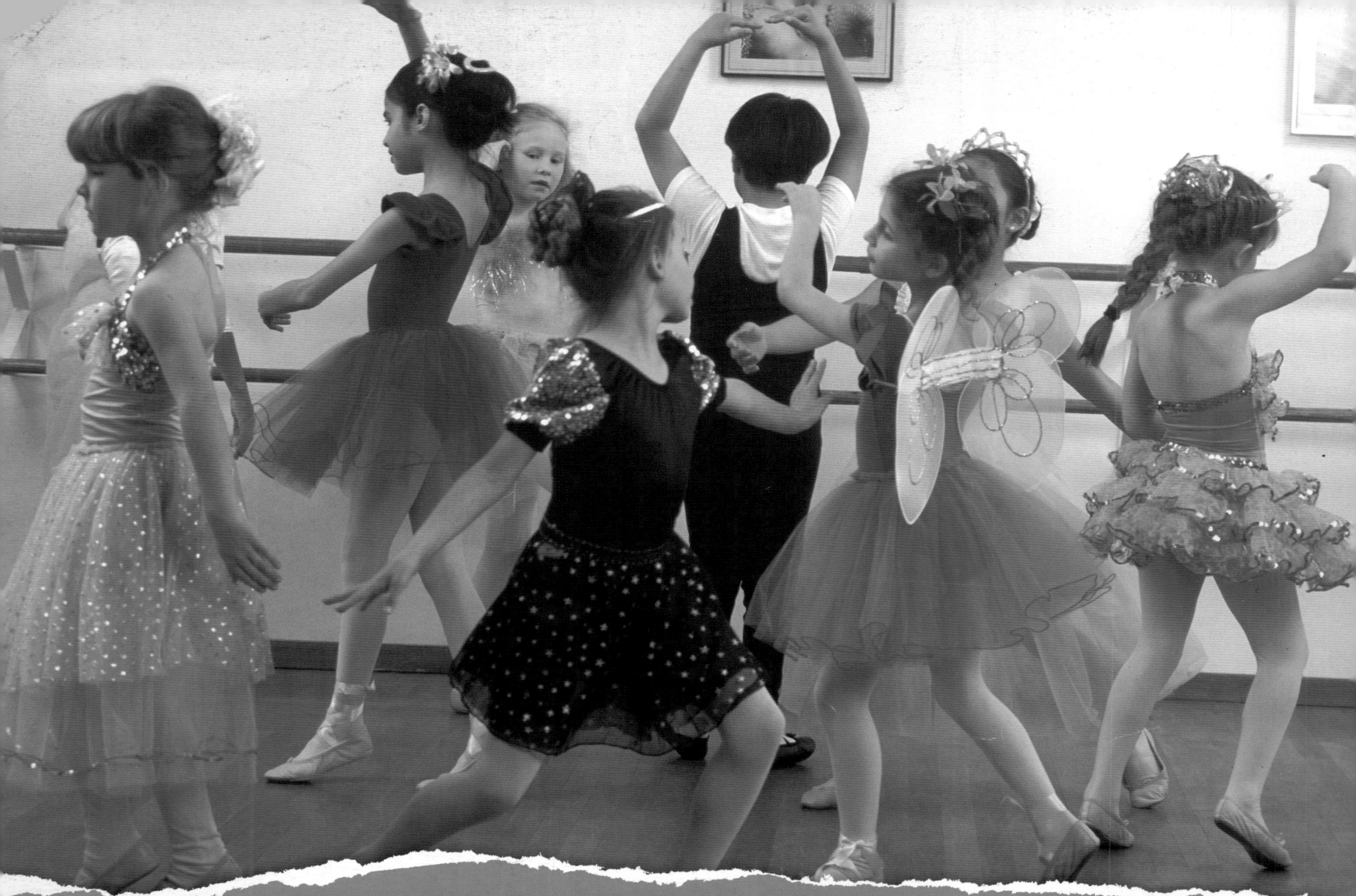

Dancing is art.

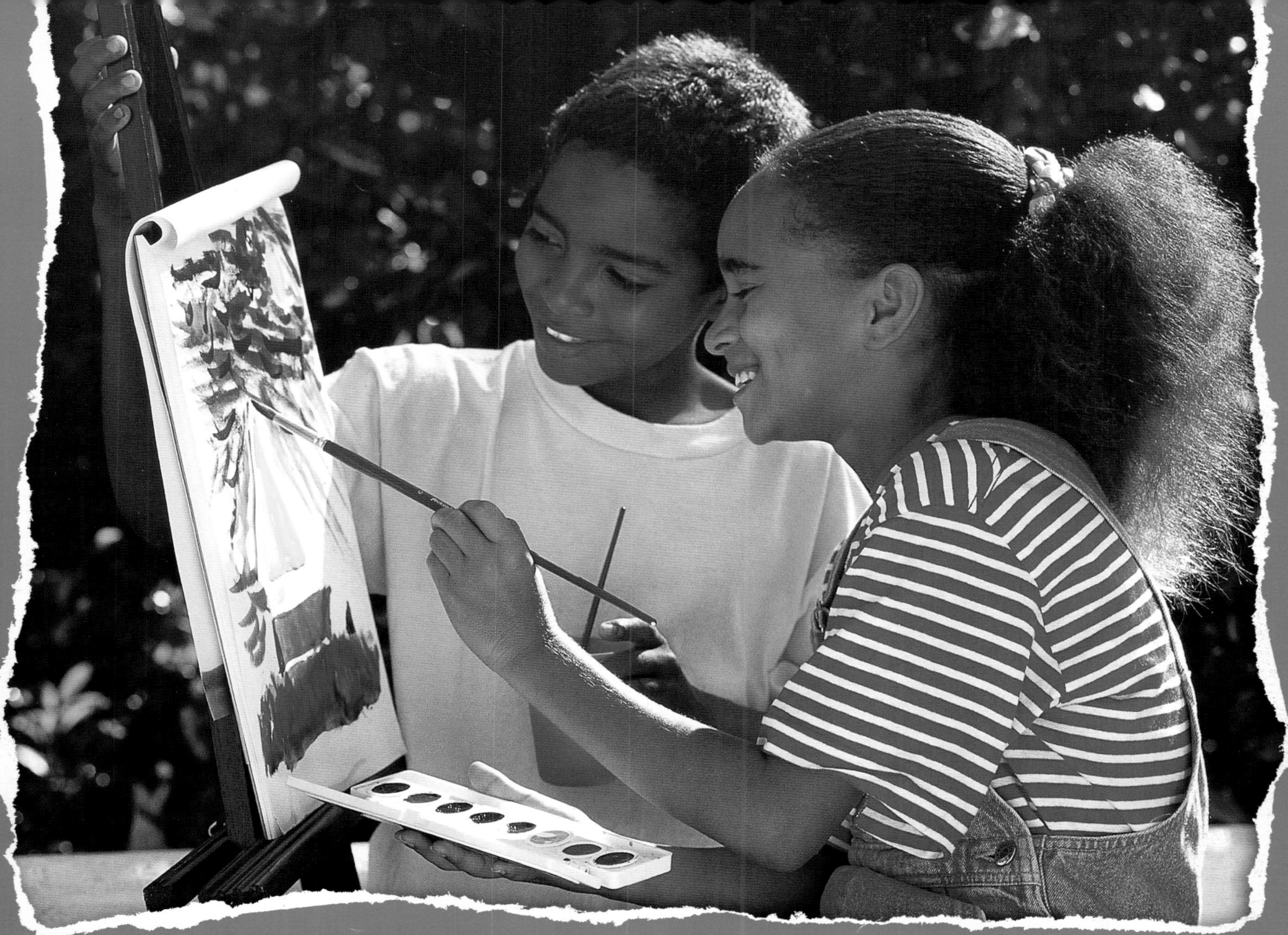

Painting is art.

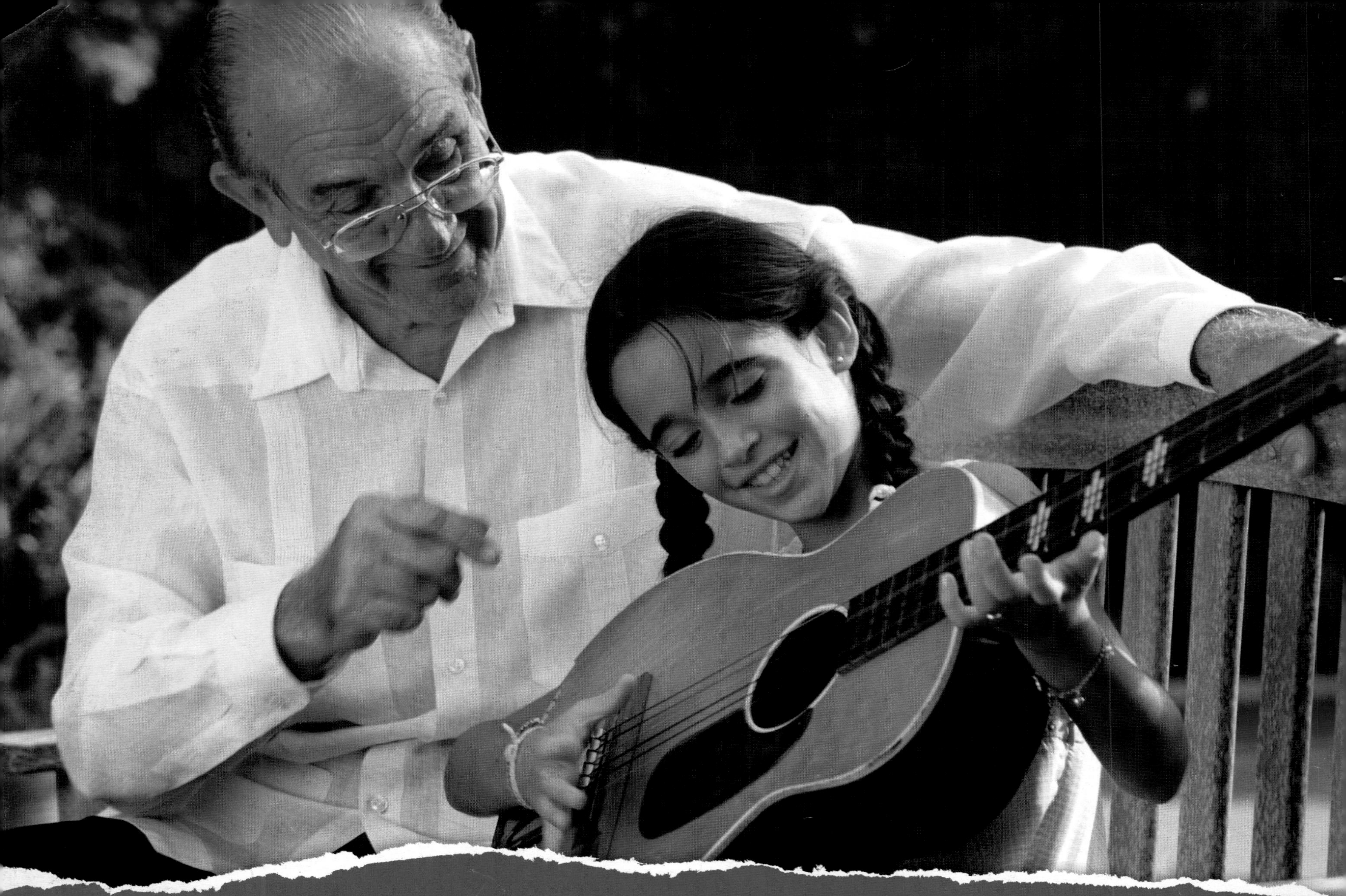

Music is art.

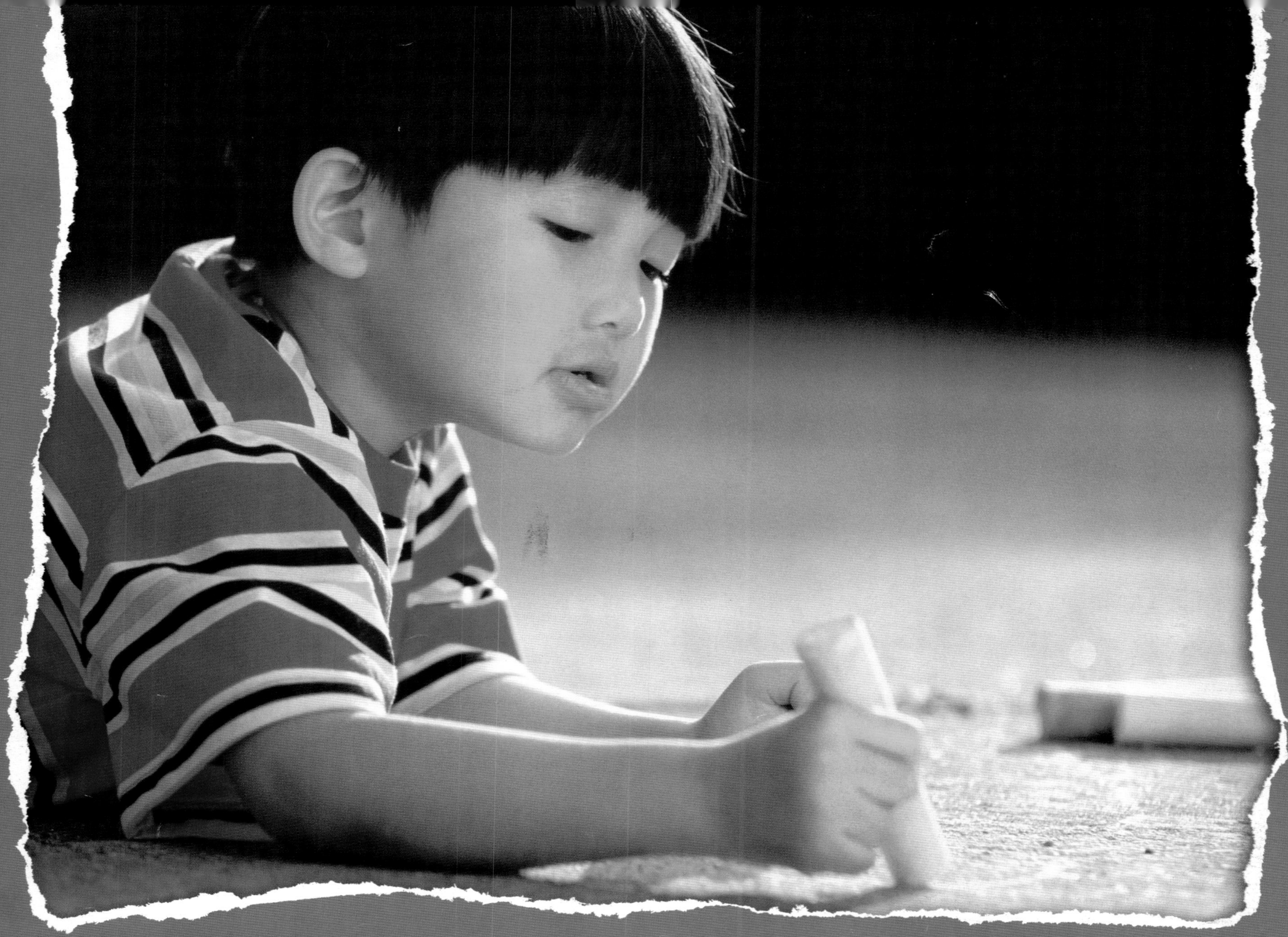

Drawing is art.

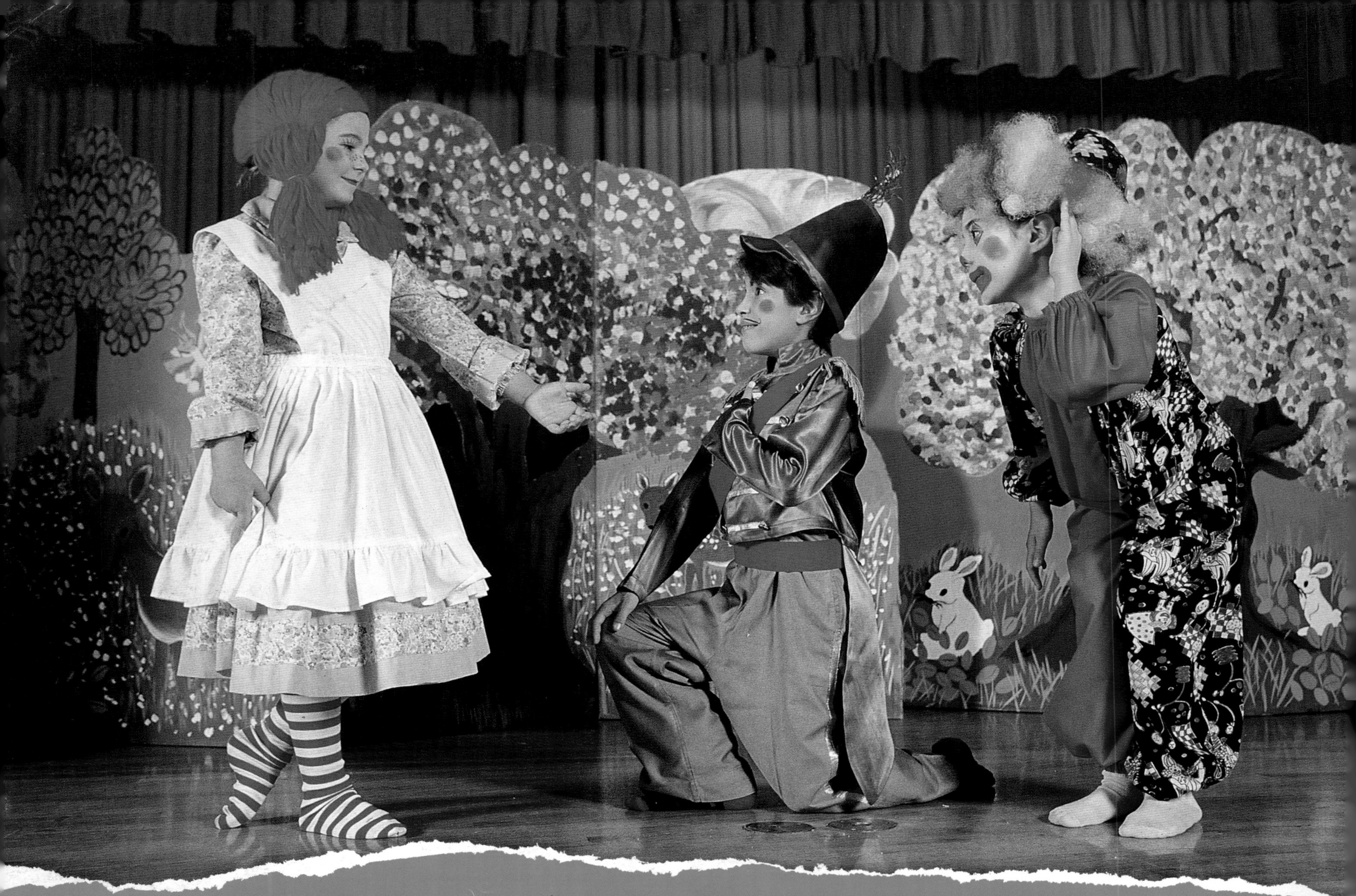

Acting is art.

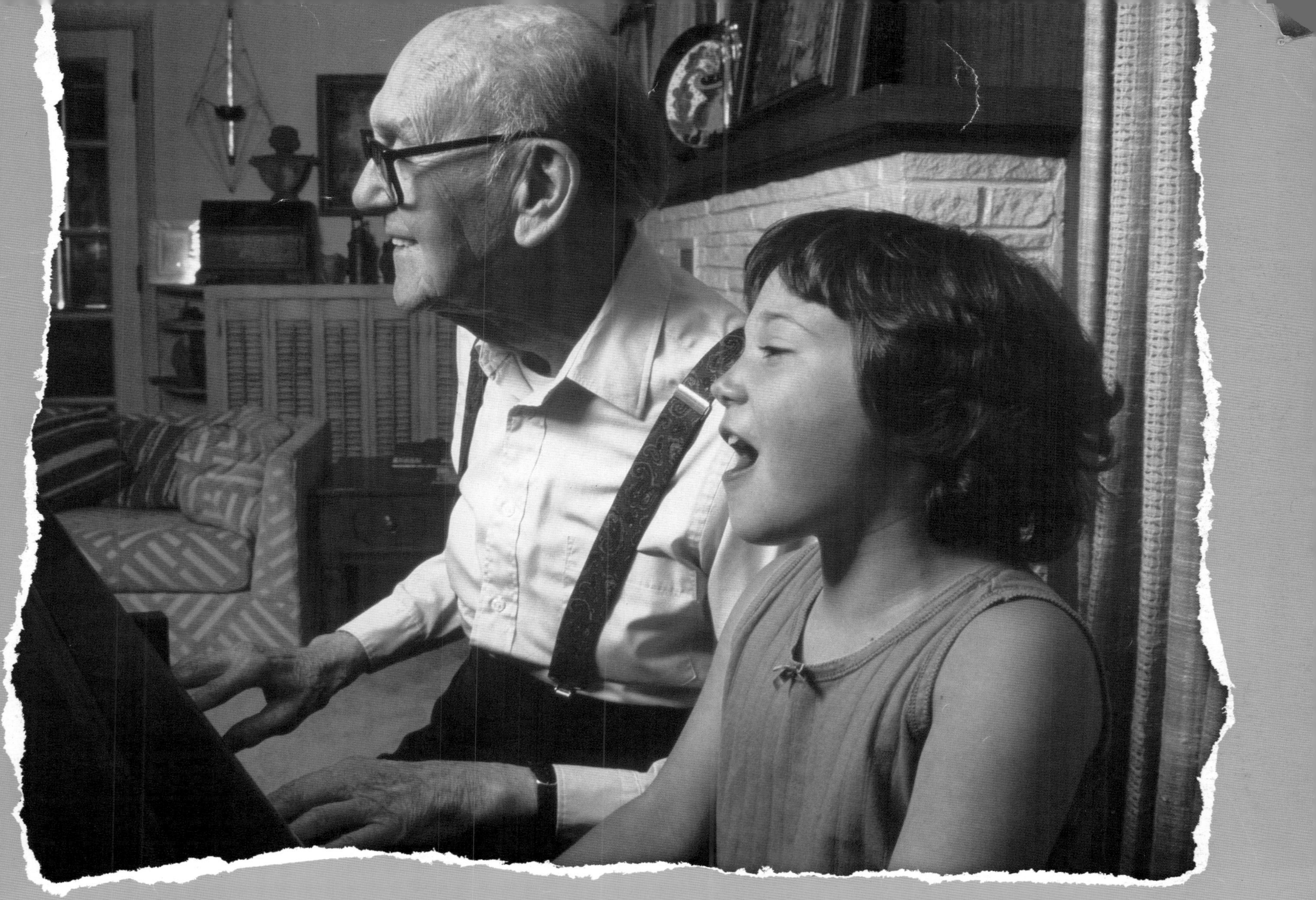

Singing is art.

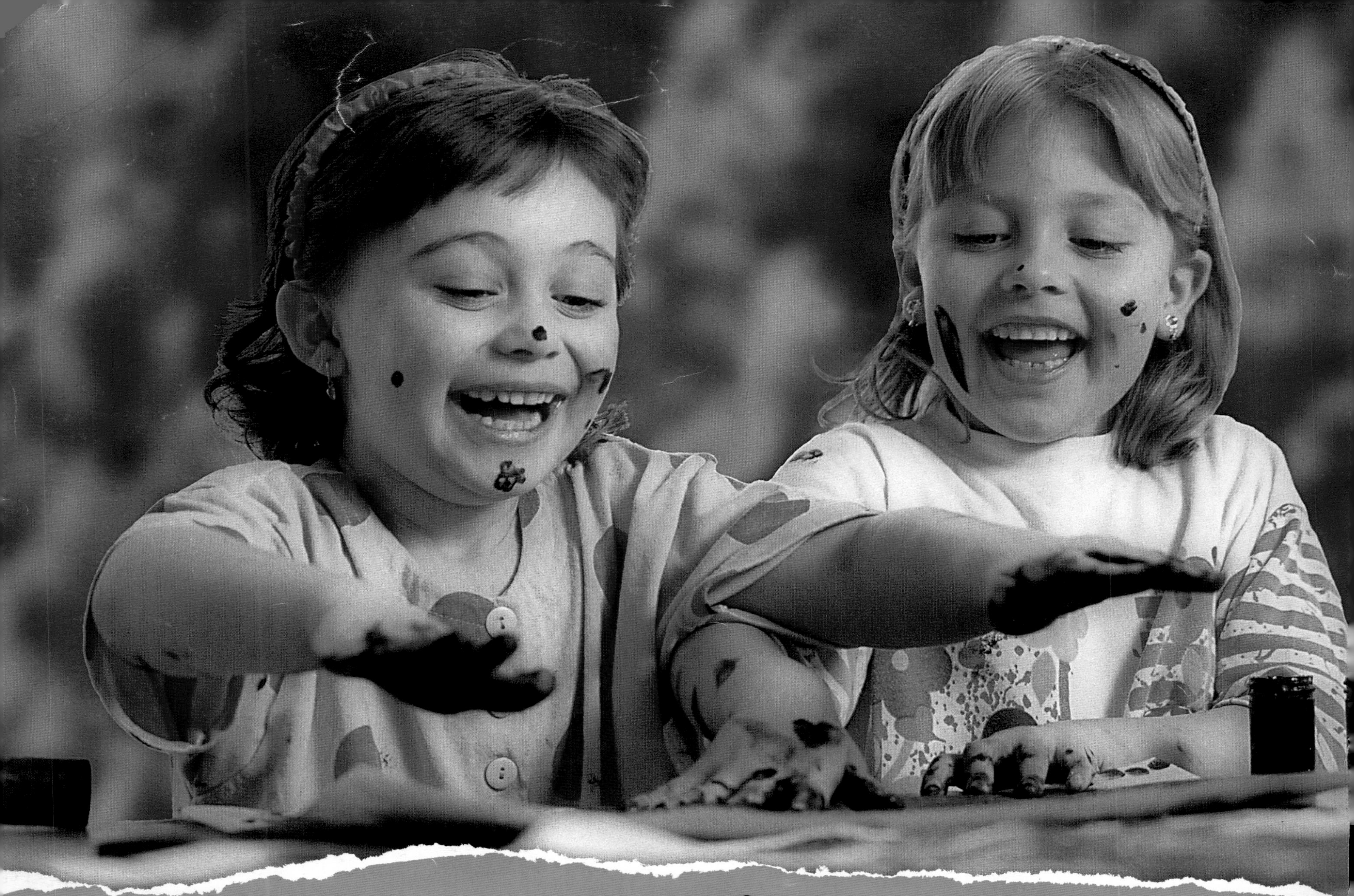

Art is fun!